I0797709

# WATER

Amy C. Rea

DiscoverRoo
An Imprint of Pop!
popbooksonline.com

abdobooks.com

Published by Pop!, a division of ABDO, PO Box 398166, Minneapolis, Minnesota 55439. 

Printed in the United States of America, North Mankato, Minnesota.

102019
012020

THIS BOOK CONTAINS RECYCLED MATERIALS

Cover Photo: Shutterstock Images
Interior Photos: Shutterstock Images, 1, 5, 7, 15; iStockphoto, 6, 8–9, 10–11, 13, 14, 16, 17, 19, 21, 22, 23, 25, 26, 28, 29, 30, 31; Xinhua/Alamy, 20; Kevin Ebi/Alamy, 27

Editor: Sophie Geister-Jones
Series Designer: Jake Slavik

**Library of Congress Control Number: 2019942486**

**Publisher's Cataloging-in-Publication Data**

Names: Rea, Amy C., author.

Title: Water / by Amy C. Rea

Description: Minneapolis, Minnesota : Pop!, 2020 | Series: Natural resources | Includes online resources and index.

Identifiers: ISBN 9781532165894 (lib. bdg.) | ISBN 9781532167218 (ebook)

Subjects: LCSH: Water--Juvenile literature. | Water-supply--Juvenile literature. | Natural resources--Juvenile literature. | Environment--Juvenile literature. | Ecology--Juvenile literature.

Classification: DDC 333.91--dc23

# WELCOME TO DiscoverRoo!

Pop open this book and you'll find QR codes loaded with information, so you can learn even more!

Scan this code* and others like it while you read, or visit the website below to make this book pop!

**popbooksonline.com/water**

*Scanning QR codes requires a web-enabled smart device with a QR code reader app and a camera.

# TABLE OF CONTENTS

# CHAPTER 1
# A SUMMER STORM

The girl stands at the edge of the farm. A field of corn stretches out before her. The soil is very dry. Dark gray clouds rumble in the sky. Bright lightning flashes. Soon rain starts falling in sheets.

WATCH A VIDEO HERE!

*Rain is a type of precipitation. Precipitation is the word for water that falls from clouds.*

Big droplets shower the plants. The water soaks into the ground.

Rain is part of the water cycle. All water on Earth is part of this pattern. The water starts as part of oceans, rivers, or lakes. Then, this liquid water **evaporates**. It becomes a gas called water vapor. The gas rises up into the air. There, it forms clouds.

*Clouds form when water vapor cools and becomes tiny water droplets.*

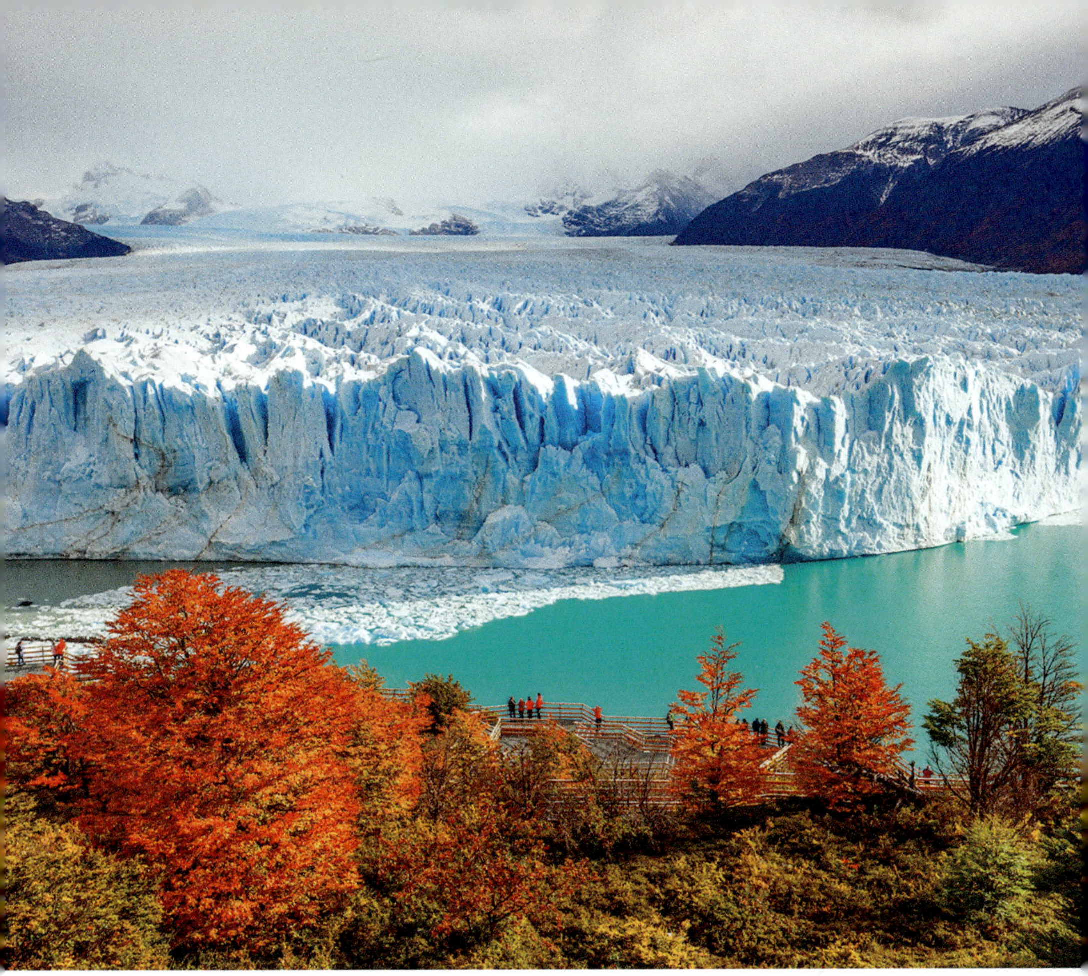

*Glaciers are large blocks of ice. They are made when large areas of snow turn into ice over time.*

**Ice is water in a solid form. Less than 2 percent of the water on Earth is frozen.**

Clouds are made of tiny drops of water. The water drops mix together. Eventually, some drops get too big and heavy to stay in the air. They fall to the ground as rain.

*Many mountain lakes are created when snow and ice melt.*

Some rain falls into the ocean. Other rain falls on land. This water sinks below the ground's surface and flows back to lakes and rivers. There, it evaporates again. This cycle continues over and over.

# THE WATER CYCLE

Rain or snow falls from clouds.

Some water flows into rivers.

Some water seeps into the ground.

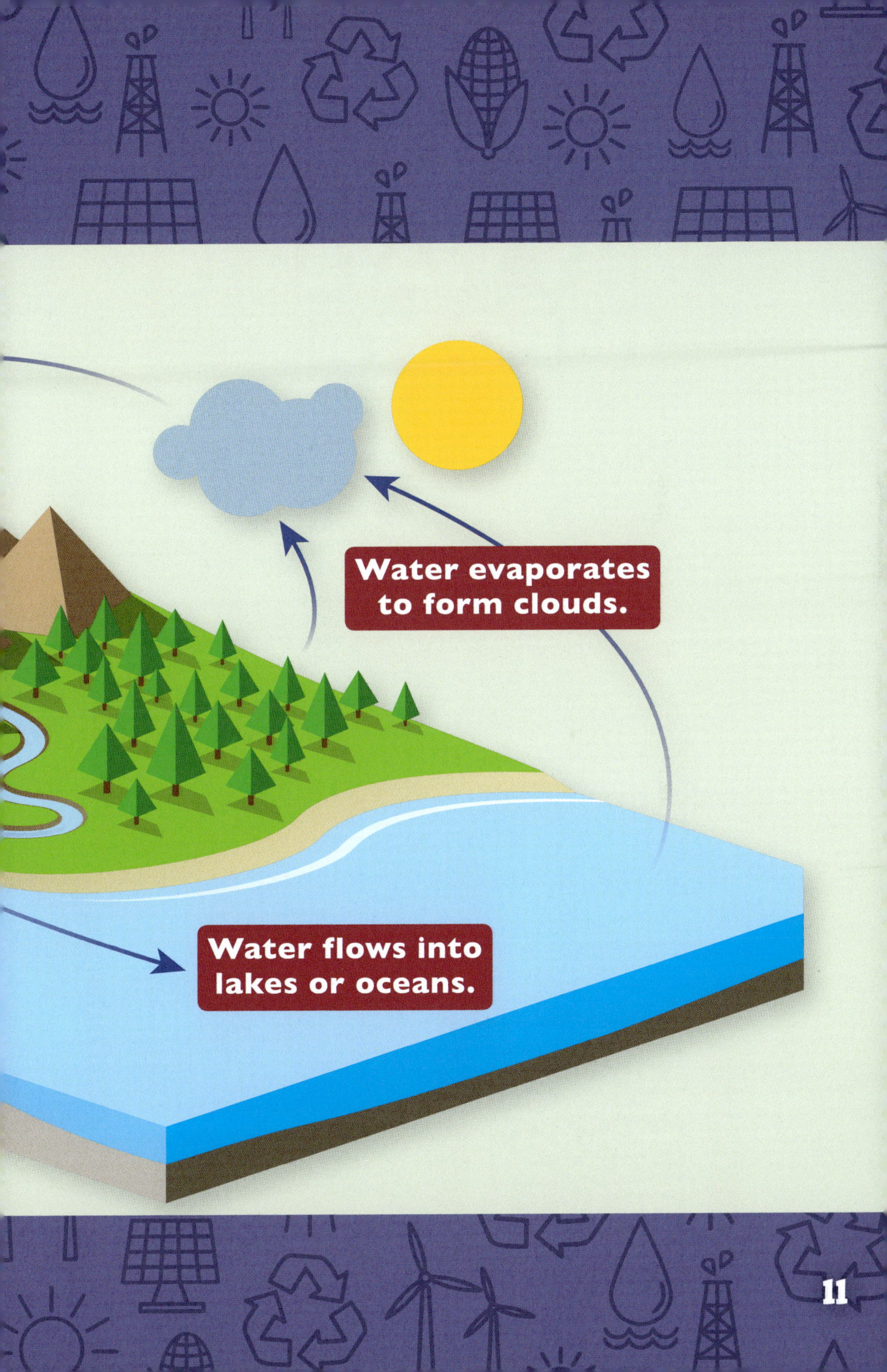
Water evaporates to form clouds.
Water flows into lakes or oceans.

CHAPTER 2

# WHY WATER IS IMPORTANT

Water covers most of Earth's surface. But most of that water is salt water. It is found in the oceans. Scientists think that more than one million different creatures live in the ocean. But not all living things

LEARN MORE HERE!

*Cows can drink up to 30 gallons (114 L) of fresh water per day.*

can live in salt water. Instead, many creatures need **fresh water** to survive.

Some fresh water is found on Earth's surface. It is part of bodies of water such as rivers and lakes. But most fresh water on Earth is underground. This **groundwater** is between soil and rocks. People dig wells to reach it.

*Lake Superior is the largest freshwater lake in the world.*

# GROUNDWATER

Groundwater can be found in many layers of the ground. Rock layers that store water are called aquifers.

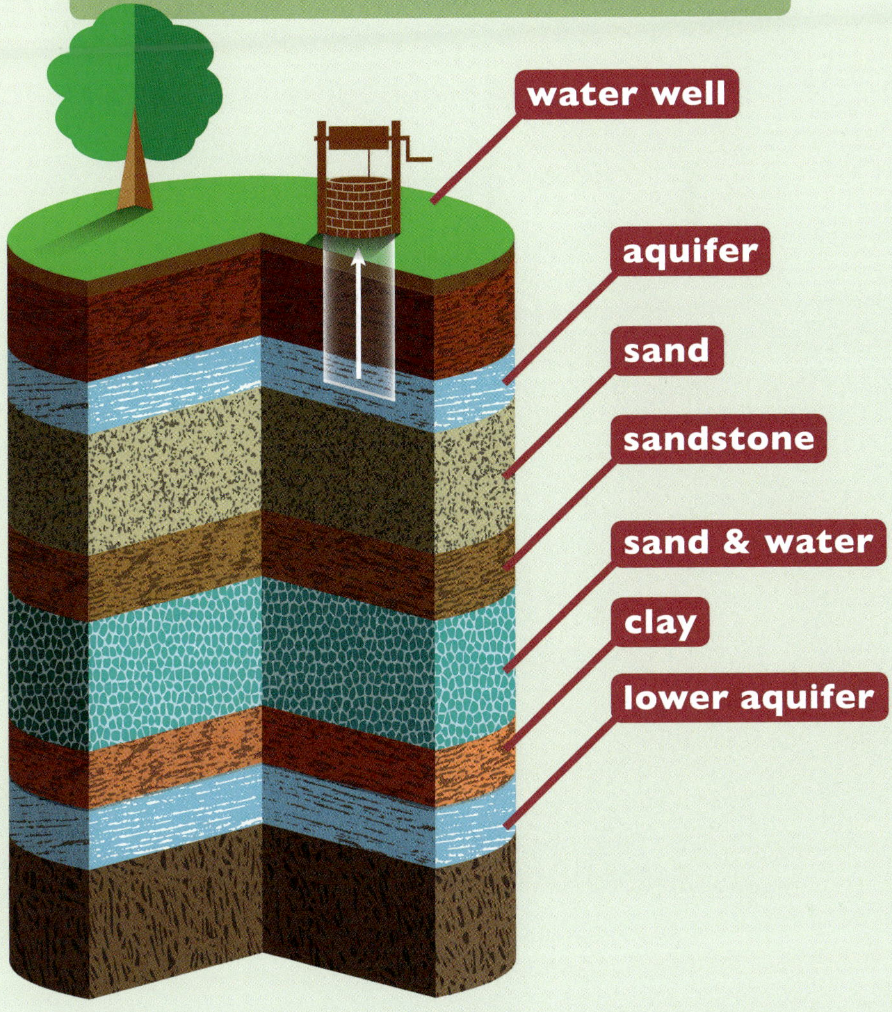

Fresh water has many uses. People use water for cleaning and cooking. It is used in **sewage** systems too. People, animals, and plants all need water to live.

*Hippopotamuses spend approximately 16 hours per day in the water.*

*Hoover Dam is one of the tallest dams in the United States.*

## MAKING WATER WORK

People can use water to produce power. For example, many **dams** generate electricity. Lots of water is forced through a small opening in the dam. This water spins a propeller. The propeller is connected to a **generator**. The generator uses the motion to produce electricity.

**DID YOU KNOW?**

**Families in the United States use about 300 gallons (1,136 L) of water every day at home.**

## CHAPTER 3

# WATER PROBLEMS

Approximately half of the people on Earth do not have ways to get the **fresh water** they need. In some areas, there is not always enough water. Or the water is dirty and makes people sick.

COMPLETE AN ACTIVITY HERE!

*Many people have to walk miles to get water.*

Scientists worry there is not enough fresh water on the planet for everyone.

**Pollution** is also a problem. Some rivers and lakes are filled with waste and chemicals created by humans. This harms the plants and animals that live there. Other chemicals get into the ground and can make water unsafe to drink.

*The Citarum River in Indonesia is one of the most polluted rivers in the world.*

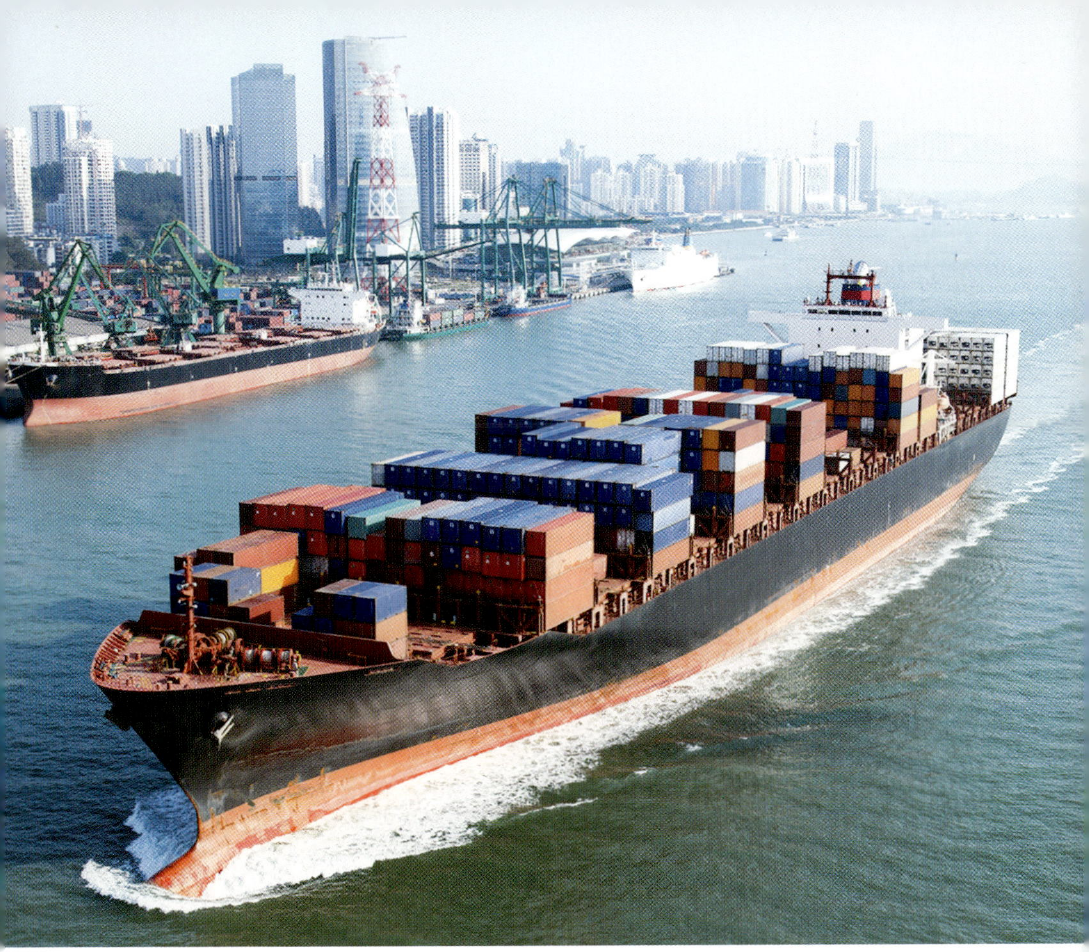

*People use ships to transport many products. These ships can leak waste or chemicals into rivers, lakes, and oceans.*

**DID YOU KNOW?**

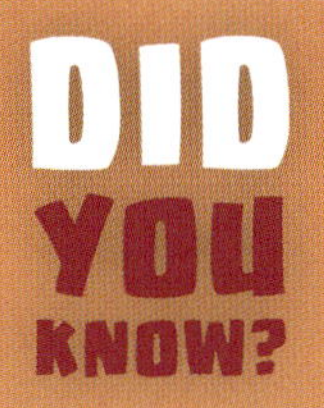

**In 1969, Ohio's Cuyahoga River was so full of chemical waste that it burst into flames.**

*Approximately eight million pieces of plastic enter the ocean every day.*

Oceans are also polluted. There is so much plastic waste in the ocean that it forms garbage patches. Scientists say one patch is more than twice the size of Texas. Animals eat the plastic waste and can be hurt or killed from it.

*Turtles can get tangled up in plastic bags or other trash.*

## CHAPTER 4

# WHAT PEOPLE CAN DO

People are working to take care of the world's water. Governments are creating laws to protect the water. Factories are trying to use less water. Organizations are cleaning the oceans.

LEARN MORE HERE!

*Scientists believe approximately 15 percent of all creatures in the world live in the ocean.*

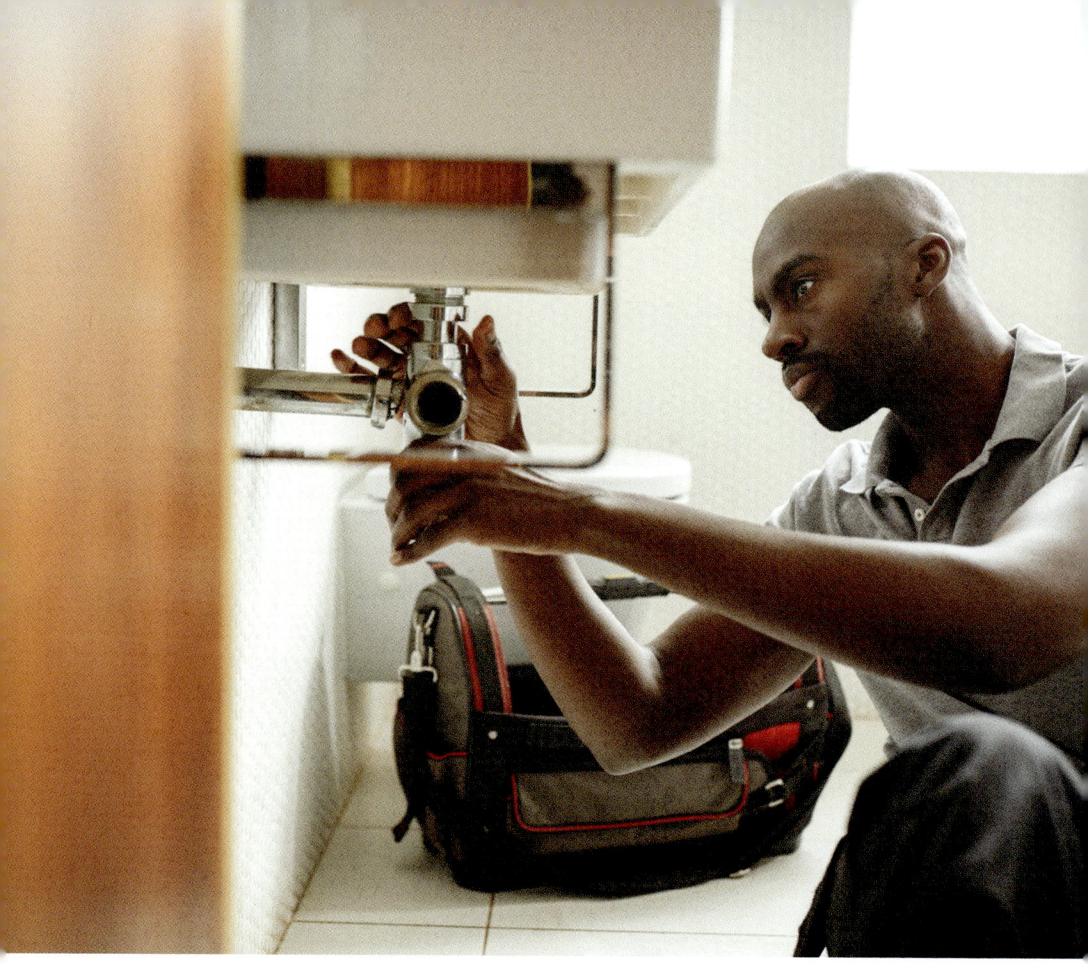

*Leaky faucets can waste thousands of gallons of water each year.*

**DID YOU KNOW?**

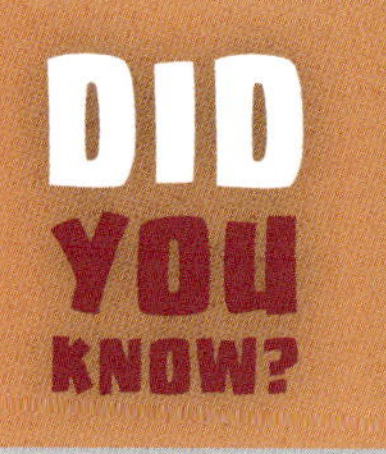

**Taking a shower usually uses less water than taking a bath.**

Families can work to save water too. People can turn off the water when brushing their teeth. They can fix leaky faucets. Families can pick up trash near storm drains and beaches.

*Storm drains in cities and towns are filled with polluted water. Drain water flows into streams, lakes, and oceans.*

*Wastewater treatment plants clean polluted water so it can be put back into rivers and streams.*

Cleaning up polluted water is not easy. It can cost a lot of money. Polluted **fresh water** can be unsafe to use for years. For example, one lake in

Minnesota has taken more than 20 years to be cleaned. But clean fresh water is important for the future of the planet.

*A scientist who studies water is called a hydrologist.*

# MAKING CONNECTIONS

## TEXT-TO-SELF

When do you use water at home or at school? Are there ways you could use less water?

## TEXT-TO-TEXT

Have you read other books about water? What new facts did you learn in this book?

## TEXT-TO-WORLD

Humans, animals, and plants all need water to live. What other things do people need water to do?

# GLOSSARY

**dam** – something that holds back water.

**evaporate** – to change from a liquid into a gas.

**fresh water** – water that is not salty.

**generator** – a machine that converts motion into electricity.

**groundwater** – water that is beneath Earth's surface.

**pollution** – harmful substances that collect in the air, water, or soil.

**sewage** – human waste that is put into sewers.

# INDEX

Scan this code* and others like it while you read, or visit the website below to make this book pop!

popbooksonline.com/water

*Scanning QR codes requires a web-enabled smart device with a QR code reader app and a camera.